D1588396

What the F*ck is 5G?

What the F*ck is 5G?

Kit Eaton

Illustrations by JoAnna Wendel

HODDER*studio*

First published in Great Britain in 2021 by Hodder Studio
An Hachette UK company

1

Copyright © Kit Eaton 2021

A CIP catalogue record for this title is available from the British Library

Hardback ISBN 9781529350883
eBook ISBN 9781529350890

Typeset in Celeste by Hewer Text UK Ltd, Edinburgh
Printed and bound in Great Britain by Clays Ltd, Elcograf S.p.A.

Hodder & Stoughton policy is to use papers that are natural, renewable
and recyclable products and made from wood grown in sustainable
forests. The logging and manufacturing processes are expected to
conform to the environmental regulations of the country of origin.

Hodder & Stoughton Ltd
Carmelite House
50 Victoria Embankment
London EC4Y 0DZ

www.Hodder-Studio.com

Introduction

In the beginning, there was darkness and deeply un-Instagrammable desperation. Hollow winds blew sadly through the telephonic emptiness, and among the people there was a great angry gnashing of teeth and much sorrow as slow Internet download speeds and patchy network coverage blighted the shadowy land.

And then a voice came out of the darkness and it spaketh . . . spoketh . . . er . . . said quietly, 'Despair no more. Get ready. A one, a two. A one, two, three, four . . .' Then it stepped up a gear, turned up the bass and in a deep, rumbling shout that shook the world of high-speed mobile communications, it concluded:

5 . . . G!

And lo, there was fanfare and someone did turn the dancing spotlights on and the laser show, and cannons did fire and also giant fireworks did streak brightly across the sky because the special-effects team had been given a massive budget and they were damn well going to burn through it with style.

Yet in time, when the noise and tumult had died down and enough minutes had passed for video clips of the fireworks to be slowly shared upon social media, the people watching the show did stop and think. They did mutter among themselves, and there was much confusion. And they did raise their voices up and say, 'Oh yes. That's all very nice.

THE LAND BEFORE INSTAGRAM

Great! But . . . um . . . WTF is, when you get right down to it, 5G?'

*

It's a good question, isn't it? Those two little characters describe an upcoming, imminent and even already kind-of-happening mobile phone revolution, and they seem to be pretty much everywhere right now. In the media, in discussions between confused town council members, in conspiracy theories wildly shared on the Web and, most obviously, splashed across the latest smartphone and phone network adverts – with the laser shows and the fireworks and the whole shebang. 5G really is being pushed as the best thing since sliced bread, and it's clearly so, so, so much less bread-like! Sorry, I mean so much better, so much faster, so much more secure, more powerful. 5G, the ads may have you believing, is practically capable of delivering more joy to your life than mere 4G.

But if you think about it, it seems like only yesterday that 4G itself was the bee's knees when it came to smartphone technology, and we were all scrambling to buy new phones that supported this wonderful, speedy, revolutionary new breakthrough. Meanwhile, you've seen it, that little '3G' warning pops up on your device screens when the network signal is a bit weak, so surely 3G can't have come much before 4G, can it? It's still in operation, after all. No one's thrown a big switch and turned it off. Have they?

Now I come to mention it, there's another question about what that G in 3G, 4G and 5G means. And what is it about the 5 bit that merits all the media fuss and exuberant promises to change the world as we know it? To answer these questions, and more, we need to look at a bit of recent history and do a bit of reasoned thinking.

I can imagine the chill that ran down your spine at those words, but bear with me. It's worth it, and could even be really important given the complexity, promise and, oh ho ho, yes, controversy of 5G. Things may even get a bit philosophical (I mean, let's do a bit of navel-gazing, and think about the old-fashioned word 'telephone' really, really hard. It means 'sounds from far away' or something similar. Now ponder how 'telephone' relates to the ultra-high-tech slab of techno-magic that sits in your pocket. You know, that little box that serves up video and music, takes photos, amuses you during bouts of insomnia and, sometimes, lets you talk to people far away. See? Weird, isn't it?). To properly get a grip on what 5G really is, I'll also need you to grasp a few amazing scientific ideas, grapple with the merest smidgeon of maths and learn an atom or two of physics. I'll make it funny, though, to help you along the way. 'Funny physics,' you're thinking. 'Is that even possible?' It is, I promise. You'll have a positive reaction.

So let's tell the tale of 5G. A fabulous story, it will shine a bit of a spotlight on the mysteries of some very clever phone inventions, and explain how 5G may even come to influence bits of your future life that you hadn't considered.

The 5G tale starts, funnily enough, not with a five or even a one, but a big, confusing zero . . .

Chapter 1

Zero G to 4G: a Brief History of the Gees

No, not as in astronauts mucking about in zero gravity. Zero G as in 0G, the zeroth generation, and 4G as in fourth generation. 'To understand where we're going, you've got to understand where we've come from' . . . or something like that anyway; there's definitely a fable or metaphor that's relevant here. The point is that before we get to 5G, it makes sense to look at what came before it.

So hold on to your hats – this is going to be a whistle-stop tour of all the Gees that came before the one you really want to hear about: the fifth one.

0G: phone history in the making

Back in the era just before the light bulb, and a few decades before the invention of sliced bread, several different people around the world had an amazing idea: let's create an almost magical machine that can send your voice signals a long way, in real time, over a wire! Alexander Graham Bell famously got the patent in 1876, but the thing called 'telephone' was really invented in fits and starts over many years. Then, for about a hundred more years after Bell, it was tweaked, improved, modernised, digitised . . . but one feature of the phone would remain – that wire.

Funnily enough, a weird 'wireless' technology called radio was being discovered about the same time as the phone was being perfected. This was also an incredibly magical way to transmit voice signals over a long distance. The two technologies eventually combined, and the radio telephone was born, and with it the ability to ditch that pesky wire. But radio phones were relatively rare and specialised. Lots of different systems were tried for years, right up until the middle of the twentieth century. The problem was that battery, electronic and radio technologies of the time weren't very advanced, and to our twenty-first century eyes these radio phones were hilariously massive and incredibly limited in what they could do.

This era of mobile technology was so very old-fashioned compared to what we use today that we now call it the 'zeroth generation' of mobile phone tech. It was barely mobile, really, and these radio devices were hardly phones. Nonetheless, it is where this story began.

PREHISTORIC TEXTING

TEXTING OF THE PAST

1G: when phones went mobile

The first generation, 1G, of what you could call modern mobile phones arrived in the 1970s and early 1980s, alongside the disco era of the Bee Gees.

The innovation that made 1G so different from 0G was cellular radio – which you've probably now guessed is where 'cell phone' comes from. Previously, radio phones relied on large centralised radio antennas, which meant you couldn't stray far from that big mast. Cellular radio, by contrast, has lots of smaller antennas connected together in a network across a wide area, even as big as a country, so a handset can move from the radio area of one to another while staying connected. Clever, eh?

Think of 0G as being a bit like lots of people in one disco dancing to the same Bee Gees record but only being able to hear it when they were near the speakers, while 1G is a bit like lots of people walking around with Walkmans listening to 1980s music no matter where they were.

Just as the radio tech was getting cleverer, so were those other areas of technology that had limited the zeroth generation of phones. There were big advances in transistor electronics and battery chemistry, which meant that 1G mobile phones were smaller and had bigger range than their 0G ancestors. Smaller is a relative term, of course – to your eyes now they'd seem comically big and awkward. They were so clunky and heavy that mobiles were frequently found in cars long before they were small enough to be a standard extra for the late-1980s yuppie, alongside shoulder pads and, yes, Walkmans.

I keep bringing up music for one reason in particular (other than being obsessed with disco, of course). It's a good way of introducing the idea of analog technology (versus the numerical digital tech you know nowadays). Your voice, for example, is an analog signal as it moves through the air – it is made of smoothly changing waves of energy. Vinyl records stored these waves as smooth-shaped grooves in the plastic, and analog cassette tapes recorded sound waves as smoothly-changing magnetic fields on magnetised tape. At the 'playback' end of a record deck or tape machine, the groove shapes or magnetic signals were read and turned into smooth electrical signals and then sound waves again.

0G and 1G telephones used analog technology by converting sounds into 'smoothly changing' electrical signals, processing those and sending them out as radio signals that, in a way, matched the shapes of the original sound waves. At the receiving end, these radio signals were read and processed into electrical signals, then back into sounds at frequencies our human ears can hear.

So while 1G cell phones were cool, modern and revolutionary because they were now, truly, portable, the radio tech that made them work was a little old-fashioned. Analog technology can be artistic and wonderful (just look at the nerd value of vinyl records nowadays!), but as more and more people wanted to use mobile phones, this way of sending voice over radio was simply not up to the job. Why? Imagine it a bit like this: two tin cans and a bit of string *kind* of work for transmitting analog voice messages across a short distance. But when you have hundreds of cans and string stretched across a room, you run out of actual space for the string to go – and then it's time to try something different. Now, this is actually a terrible metaphor for how 1G phones worked, and the 'space' that was taken up by all these phone connections was radio space in the airwaves . . . but it does give you an idea of roughly what was going on!

2G: a farewell to analog

CDs, a digital sound system, were launched in the 1980s, and before long the CD was overtaking analog vinyl and cassettes. This was firstly because they were smaller than records, and secondly because they could store a lot more information in comparison with vinyl and tape, due to the fact that the method of storage had changed. Instead of those beautiful smooth, analog grooves, CDs stored their information as chunks of digital code; ones and zeroes we know as 'binary'.

Around the same time, 2G phone networks were launched, and they used this same digital technology. Instead of sending a smooth-changing radio version of a voice signal, 2G handsets sent and received digital or 'coded' versions of the sound waves over the radio network – those little blips and bloops of radio that make up the ones and zeros of computer data. Think of it as being a bit like using a number to approximately describe the shape of the sound waves, instead of accurately recreating every detail of the waves in analog form.

Just as with the physical space on CDs, digital audio signals now took up much less radio 'space', so more phones could connect to the same radio mast at the same time. Another advantage was security – because it was now a numbers game, the signals could be encrypted with some clever maths, so only the user and sender could hear the call.

The last advantage was probably the biggest, in hindsight. Since sounds were being converted and sent over the radio as numbers, non-voice information could easily be sent as numbers too. Someone thought it would be handy to have a 'short message service' so that users could send text-based communications to other phones. It took a little while, but when this became popular, teens the world over rejoiced: the SMS, or text message, was born. Before long, billions of SMSs were flitting their way across the globe, and texts – and, later, multimedia messages – set an important direction for the future of the phone: less talking, more typing.

The 2G mobile phone system also included more international protocols, so that the way phones worked around the world, and even inside one country, was more standardised (not *completely*, though; thanks to disagreements, there were several competing standards). This meant the phone industry could concentrate on making fewer different types of phones and network equipment. Plus consumers had, in theory, more choice.

2-and-a-bit G

Society changed during the 1990s: shoulder pads were out, boy bands were in, and cell phones became more commonly used around the world. 2G itself changed and improved. One technological leap, the General Packet Radio Service (that GPRS corner icon that nineties kids will be so familiar with), gave cell phones new tricks like multimedia messages and basic Internet connection. GPRS basically chops the signals going to and from the phone into little lumps or packets, versus the older way of connecting, which was like an old-fashioned phone, with a permanent 'circuit' when it was connected. Think of the older system as being like each phone having its own 'wire' to send signals over, whereas GPRS sends little pieces of the signal spread out in time, interspersed with the chunks coming from other phones. GPRS allowed more connections at once because lots of phones' packets could be jumbled together in the circuits of just one mast. It allowed fast data to be sent too. Anyway, GPRS was when the mobile Internet arrived! Looking back, it was really the 'very slow and functionally crippled mobile Internet'. But it did work.

EDGE tech (Enhanced Data rates for GSM Evolution . . . you don't need to know exactly how this worked, but the important bit is 'faster data') was the next big advance, arriving in 2003. EDGE worked on the core hardware of GPRS (so the phone networks didn't need expensive physical upgrades) but added cleverer software inside the system to extend what GPRS could do. With EDGE's improved audio and signal coding tricks, more data could travel over the network and get to and from a compatible cell phone more quickly. It was basically a little bit like 3G . . . without quite being 3G.

'More data, faster . . . hmm, interesting. And they all want more?' This, you may imagine, is the kind of thought that went through phone network CEOs' minds at the time. Detecting the public's thirst for data and speed, the networks began a race to sell us faster data access for more money.

3G: the first G you probably heard of

3G networks first went commercially live in late 2001, in a very small number of places. They used all sorts of different technologies, but basically built on the digital strengths of 2G by including even cleverer digital coding processing into the systems that make a cell phone network. 3G networks also used different radio technologies than 2G did – including a bigger range of frequencies. I'll go into the advantages of this later, but suffice to say that when we're talking about networks and phones, more is always better. None of that 'less is more' business!

You'll have spotted that the dates for 3G mean there was an overlap with networks using GPRS and EDGE systems. The industry was trying out all sorts of tricks and upgrades, and 'full' 3G took absolutely ages to arrive in some countries for all sorts of reasons, including questions about the cost-effectiveness of 3G and the fact that it used more radio frequencies than before. Some networks stuck with GPRS and EDGE as their main system until the late noughties . . . and they are still switched on in plenty of places. Nothing's ever simple in the phone industry, it seems!

By using new radio systems, better security, different electronics and different encoding for the digital signals, 3G networks could achieve much faster data speeds. We're talking more than one hundred kilobits of data per second. At the time, that was incredibly fast, but to put it in perspective, it would take *thousands* of seconds to download a 1GB file. This means it took longer to download a file like a medium resolution movie than you'd take to watch it! That wasn't really a problem back then because the phone display tech available couldn't really do justice to video – it was just torture on the eyeballs.

3G's speeds were plenty fast enough for short video content and images, though, and they really allowed mobile data technology to flourish and for consumers of all types to get used to data everywhere. We're not *quite* talking the era of mobile meme videos, but perhaps the start of this sort of idea. Certainly enough data, though, for people to engage in online arguments while waiting for their bus to work about whether GIF is pronounced *giff* or *jiff*.

Thanks to 3G, the smartphone era had arrived. Remember BlackBerrys? This was *their* time. If you *don't* remember, these clunky devices were a revolutionary early smartphone, mainly aimed at bringing email to business people. Fiddly and expensive at first, they had tiny QWERTY keyboards and basic apps like email, calendar, a browser and – gosh – games! In the early 3G years, sales of BlackBerrys and other similar data-centric consumer and business 'smartphones' exploded around the world, as bosses everywhere realised they could now email their employees at home at 7 a.m. with a new project, then again at 7.30 a.m. with a message saying, *What progress have you made?*

Of course, another advantage of 3G was that it was supposed to be a more international system than 2G – people around the world were used to cell phones by now, and governments were realising there were advantages in standardising the things.

*This actually happened to me the very first morning I had my first BlackBerry. The soaring joys and deep diffi-culties of being *always* connected were revealed to me all of a sudden that day.

But in the manner of committees everywhere, there were disagreements between governments, regulating authorities and telecoms giants about *which* 3G standards to follow and which radio frequencies could be used in different countries. So roaming to another country was sometimes possible, if you had the right phone and paid for the privilege.

But standardisation didn't extend to everything. Phones had different screen shapes and sizes, manufacturers had different ideas about operating systems, headphones and charger cables. Phone networks even had different 3G speeds, and some phones wouldn't work on another network at all, *ever*, so if you switched providers you might need to buy a new phone.

I don't want to blow Apple's trumpet for them (they have a very expensive and clever PR department that does that job *extremely* well), but I'll give it a little toot here. A seminal moment in the 3G era was the arrival in 2008 of the iPhone 3G. Compared to the awkward small-screen QWERTY keyboard smartphones and the thousands of 'stupid' phones that came before it, this device was a quantum leap in usefulness. It really made the most of 3G mobile data, with a large screen, proper Internet, and apps – mobile videos like YouTube, for example, finally made sense! While the original 2007 2G iPhone was a very limited release, the iPhone 3G went on sale in twenty-two nations and was compatible with many different phone networks across the world. It can almost be argued that it set us all on the path to 5G. But not before the phone industry tried out a few other things . . .

THE AGE OF THE SMARTPHONE BEGINS.

3-and-a-bit G

Like 2G, 3G services were tweaked and improved over time to make them better, faster, stronger. New technologies like HSPA, HSUPA, HSPA+ arrived. Forget the deeply uninteresting meanings behind these acronyms; you just need to know that while they relied on more or less the same radio and computer tech as earlier systems (so that, as with the 2G EDGE upgrades, phone networks didn't have to install vast amounts of new equipment), they could deliver much faster data rates to and from users' phones. We're talking super-speed data at several megabits per second! That's ten times faster than 2G EDGE, so on a good 3G connection a 1GB movie could now take less than an hour to download. Painfully slow, but since a movie lasts longer than an hour, not all that bad.

Some of these new speedy technologies tempted phone networks to say, 'Look, forget 3G . . . we have 4G now!' and to try to get consumers to pay for the privilege of what was merely 3-and-a-bit G. One of the technologies involved in this was called LTE, for 'Long Term Evolution'.

Of course at this time the proper 4G standard was being carefully drawn up to solve exactly this sort of problem.

4G: the need for speed

By the time 4G rolled around, there was a concerted effort by different standards-setting bodies around the world to agree on how next-gen phone networks should work. The idea was that this would be better for phone manufacturers, who would have to make fewer devices to meet the different standards; and easier for consumers, who could take their phones from place to place around the world. That was the hope, though reality turned out a little differently. The 3-and-a-bit G muddle with LTE shows how this standards effort didn't quite work at first, which is probably no surprise to anyone who's ever sat on a committee.

As expected, 4G, which arrived in phases from around 2010, used *even more* radio frequencies and more bandwidth* than 3G and offered even faster data. As well as faster data, the idea was that 4G networks would be more reliable when handing a connection from one radio cell to another, and that more users could connect at a time. The speed goal was 100 megabits per second for people on the move, like pedestrians or commuters; and a maximum of 1 gigabit per second for stationary connections. These speeds translate respectively to around 80 seconds and 8 seconds

---

*Don't get hung up on the words 'frequencies' and 'bandwidth' here. I'll explain the science of all this in the next chapter.

to download a 1GB file – much less time than you need to watch a movie!

This is likely the tech in your phone right now, so you'll know that in real life the network often isn't that speedy. When you have an excellent signal, it's sometimes fast enough that it roughly compares to home broadband and can allow for real-time streaming of large files, or for you to comfortably download HD movie files much bigger than 1GB. But when you have a dodgy signal, 4G can be pretty sluggish. Some nations' networks deliver a typical user around 2 to 10 megabits per second, many, many times slower than the promised maximum, and actually not much different to 3G. 'It's a scam!' I hear you muttering to yourself as that hilarious video of a kitten tipping milk all over the floor pauses and buffers for the twelfthtieth time. It's not entirely a scam, I promise. The truth is that the results *do* vary – in some places, depending on the phone network and tech used, things go much faster, at maybe 20 megabits per second.

Despite its problems, 4G enabled mobile video streaming services like Zoom, an explosion in mobile gaming, video calling, and just about all of those other things we take for granted that would have seemed impossible just ten years ago. But to prevent consumers getting too pissed off, phone networks around the world recently rolled out 'improved' 4G.

Where have we heard this before?

4G+: adding some shenanigans

Just like with the switch from 3G to 4G, with 5G there have been some 4-plus-a-bit G shenanigans by some phone networks. In the US, for example, AT&T launched an improved technology based on

LTE that was genuinely speedier than basic 4G, and branded it '5GE' – though it was not, strictly speaking, 5G! Advertising officials knocked that idea on the head as being deceptive. But AT&T said it would keep a '5GE' logo on the screen of capable phones anyway, to, you know, inform the public.

*

Ignore all these shenanigans, though. If you're observant, you'll have noticed that in the journey from oG to 4G, there are a handful of big trends.

One: each generation of mobile phones incorporates such different technology that the phone networks have to change their equipment, which means you have to buy a new phone with the right sort of new electronic gubbins – great for the phone companies, phone networks and banks. Not so lovely for your actual bank balance (though at least you can sorrowfully check your diminishing funds via your banking app on your lovely 4G phone). The 'old' networks often weren't turned off though, and were left on in some places for many years.

Two: as the Gs advanced, voice communication became less and less important compared to all the other 'smart' things your phone could do. Just talk to a Millennial about making an actual voice phone call, and watch them shudder in horror.

Three: as mobile data became faster over time, and we all realised how neat it is to browse Facebook on the train, post Instagrams from the coffee shop and play games while perched on the toilet, we wanted more of it. Data, that is. Much more of it. More data, faster data; data, data everywhere.

Four: as the technology has advanced, there seems to have been more of an effort by government bodies and tech companies around the world to improve standardisation. It doesn't always work, but it is the goal.

Five: every G brought more and more users on to the network, eventually maxing out the free radio space or the tech needed to keep all that data flowing to more people, leading to the need for the next G!

Now, notice that the new Gs tended to arrive roughly ten years apart, plot these five trends beyond 4G, and you get to . . . the next G! 5G! For which you'll need another new phone, which you'll use for many more purposes than you used your 4G phone for, and it will be faster than you ever dreamed possible.

Chapter 2

What's So Special About 5G?

5G can reshape business! 5G can connect the world! 5G is the Internet of Everything! 5G will generate trillions of dollars/euros/yen/pounds in new industries we can't even imagine yet! 5G will change education, factories, hospitals, gaming . . . 5G will revolutionise cars, maybe transport in general! 5G can even change your sex life!

I made that last bit up, but it's not that much of a joke, because while there is a lot of wild stuff like this written about 5G by advertising executives who may have had one too many espressos at lunchtime, 5G is an incredibly powerful invention. And believe it or not, it really does have the power to penetrate (ahem) deeply into your life . . . But before I get to what 5G can do for you, let's look at what sets it apart from all the other Gs.

Whoosh!

The biggest thing 5G has going for it is speed. How much speed, you ask? The people who drew up the rules for 5G say it should easily reach 10 gigabits per second. Which is 10 billion bits. Whoosh indeed!

In real-world terms, that's around ten times faster than the idealised maximum speeds that were expected for 4G. Ten times faster for anything is quite impressive.

In the previous chapter, I mentioned that a 1GB file is roughly equal to a full-length movie at medium resolution; the kind of thing you download from Netflix to your phone before you get on a long flight. It's a useful example. At 5G's 10 gigabit speeds, it would take less than 1 second to download that file. You'd be able to download nearly 4,000 of them in one hour. Whoosh (again)!

The really wild thing is that 5G's top speed limit is not fixed in stone yet, and it's possible that some 5G networks could quite quickly achieve 20 gigabits per second or more. Theoretically, 50 gigabit download speeds may even be possible, which is 25,000 times more speedy than the 4G speeds some users currently experience on slow networks.

Will 5G networks reach these speeds tomorrow? From the varied and unreliable history of 4G data speeds, I would predict no, of course not. Some early 5G networks that have already rolled out have made headlines by actually delivering equivalent or only slightly faster speeds than 4G networks. Oops. 5G speeds will also vary from place to place – in a mid-2020 survey of speeds for 5G networks that have already rolled out, Saudi Arabia topped the list with 144 megabit download speeds, Canada had around 90 megabit speed and the US and UK came in last at around 33 megabits. Not whoosh at all. On the other hand, in November 2020, Nokia revealed that it had reached 8 *giga*bit speeds on a commercial network in Finland which, since a 'giga something' is a *thousand times* a 'mega something', is definitely whoosh.

Of course these are early 5G networks, and we know from experience that the real speeds often don't quite live up to the promise, but let's imagine enough time has passed and many 5G networks finally deliver those smokin' 10 gigabit speeds. Why might this be useful in tomorrow's world?

For one, file sizes are likely to climb, particularly for content like high-definition movies. The trend is moving from 4K to 8K TVs in our homes, and you may be thinking, 'Surely no one wants to watch an 8K movie on their phone?!' But I say it's inevitable, and those files are bloody massive! As in perhaps 10 terabytes for a movie!

Our phones now have very powerful cameras built in, and some already shoot 4K video or high-frame-rate slow motion. You know the kind of thing – a glorious, high-resolution holiday video of waves crashing on to the beach, backlit by a gorgeous sunset. Or maybe just your friend face-planting while carrying a tray of drinks from the bar. All these video files, no matter how good, will take up precious storage space on our devices, and we'll want them backed up speedily to the Cloud ready for when we need them.

But it's not just video content that needs fast 5G signals. As OS updates get bigger, game files get bigger. 5G's higher speeds and greater network reliability will mean you're freer to just click download no matter where you are, and it will allow game developers to craft some real-time group game experiences that aren't possible right now. We're all also getting used to video calling, and 5G could really boost the typical video quality of a ten-way office Zoom meeting.

5G speed will even affect our phones when we're *not* using them. Our 4G phones spend a good amount of time lying on the table, dark and life-less, looking like they're off. But there's actually a constant to-and-fro of map data updates, backup of photos and videos, health data and emails from your phone to the Cloud, pings from social media apps and more. You don't see it, and you don't often think about it, but it keeps your phone ticking, and every aspect of this will benefit from the higher speed of 5G.

Echo . . . latency!

Dazzling data speed isn't the only amazing 5G benefit. One other advantage 5G networks have over 4G ones is latency. Essentially latency measures the delay between the cause of something and its effect. A pretty good example is to think of it as being a bit like an echo: when you stand in a large space and shout 'Knickers!', it takes time for the waves of sound to reach the walls, get reflected and then arrive back at your ears ('. . . Knickers!'). In networking terms, latency measures how long it takes for your computer to send a signal to a distant one, and for that computer to echo a signal back. Hard-core online gamers hate latency, sometimes called lag, because a slow network can make all the difference between ignominious gaming death, and shooting an opponent precisely between the eyes.

Typical 3G network latency could be about 100 or more milliseconds – a tenth of a second. This doesn't sound like much, and if you're talking about traditional voice calls, it isn't (it takes much longer for the person you're talking with to think what to say next anyway). 4G networks did a bit better, especially LTE, with ideal delays around 50 milliseconds, about half as slow as 3G. Of course, 3G and 4G networks often failed to reach these levels, and delivered much higher latency than was hoped for.

But still, think of 4G only halving the latency of 3G and you'll be amazed when you hear the goal for 5G: around one single millisecond. That's fifty times less than 4G and possibly better than the latency your home broadband serves up! It's a crazy improvement, and though in practice latency will be higher at first, these figures have been demonstrated in a lab, which means it's only a matter of time.

Latency has more implications than those for gamers (though admittedly, few feel as passionately about lag as gamers do). Video calls will benefit from lower latency – higher frame rates and fewer dropouts should become the norm, and we'll hopefully be able to say goodbye to ugly freeze-face. Some streaming apps, which need a very fast connection to the Net (like virtual reality or augmented reality ones), will really be improved by lower latency. In fact, almost any app that gets data from the Internet and relies on real-world events or real-time movements of your phone will benefit from it. Low latency becomes even more important when you're talking about moving vehicles that rely on a 5G data connection

– a delay in getting critical information to a self-driving car, even a few milliseconds, could be disastrous.

You know who else cares about latency? Businesses. For example, for some of the fastest automated buying and selling stock-market systems, latency can cause million-dollar profit variations. Now, I'm not saying Wall Street will be run from 5G smartphones, but for all sorts of auto-mated processes, be they robots in factories or connecting head office to real-time operations out in the field, businesses will love low latency.

An orgy of connections

Remember when I was talking about lots of tin-can-and-string telephones tangling up in a room? It was a metaphor for how many 1G devices filled up the radio 'space' a phone network antenna could have. Things have moved on, and now 4G digital networks are designed to cope with a lot of people using their devices at the same time in the same area without cluttering up the radio space. Essentially, each 4G cell antenna can have a certain number of phones (tablets, 4G computer dongles and so on) connected at once without the system jamming up. This number is ideally 1,000 connections per square kilometre.

A thousand devices sounds like a lot, but in real-world 2020s terms, it's often not enough, and when there's an event like a big music festival, all those drunken dancing revellers crammed into a small muddy field can overload a typical 4G radio network's powers. So network operators often have to wheel in temporary phone masts to boost their capabilities.

The problems of supporting thousands of connec-tions to each 4G phone tower aren't only concerned with revellers trying to FaceTime from Glastonbury or Burning Man; there are simply more of us humans alive every day and we're all getting a bit crammed together, particularly in cities. And it's not just things like phones and tablets that use phone signals. Lots of things you wouldn't think of, like card payment machines in your local coffee shop, ATMs and automated train ticket machines, connect to phone networks too. According to studies by McKinsey, nearly 130 new devices are connected to the Internet every second, and there are already probably more than 20 times as many Internet-connected 'things' as there are human Internet users. This is commonly known as the 'Internet of Things' (IoT).

That's why 5G is designed to allow up to one million devices per square kilometre to be connected. Can't imagine that? Picture a thousand smartphones arranged in a row, each one metre apart (that's three feet apart if you're American).

Now imagine one thousand of these rows side by side, in a huge square. And lots of those squares spread across the country. That's a crazy number of phones! But if you add in 5G tablets, 5G computers, 5G gaming consoles, 5G AR goggles and so on, those numbers add up really quickly. Then imagine all these devices working at once in a congested city area with thousands of people living and working in multistorey buildings. Now you can see why 5G has million-device densities as its goal.

*

5G builds on a hundred or so years of invention, and trial and error. It's much cleverer than 4G, but before I explain how 5G systems actually function to give all this speed, low latency and so on, I need to take you back to school . . .

Chapter 3

Physics, Yeah!

Bet you never thought you'd read those words as a headline!

Beneath the glossy advertising and the over-excited news reports both for and against 5G, there's a lot of very clever technology. It's all based on some neat science, and without the science the key bits of 5G wouldn't work. But don't worry, together, we'll get through it. And hopefully after we're through it, you'll look back and think, hmm, physics isn't that bad at all! Or maybe you won't . . . but either way, I'm going to try to make this interesting!

Getting the Hulk on the radio

One wild fact about radio waves and light waves – the things that we puny humans 'see' with – is that they are almost exactly the same thing. They are simply electromagnetic waves, which means they are magnetic waves and electric waves all tangled together and wiggling through space. Really, they're all just waves!* Mind-boggling, no? Let's boggle you a bit more then. Gamma rays (which are like cosmic rays), X-rays and ultraviolet are all the same as visible light and radio waves as well. They're just different sizes.

What distinguishes all these types of waves – radio, gamma, visible light you can see and radio

---

*For the science-minded: I'm going to skip over the notion that light is also a particle. That's wild, wonderful quantum mechanical stuff that's not needed here. Ironically, the chips inside your phone *are* designed with quantum issues in mind, otherwise they just wouldn't work.

you can't see – comes down to counting. Imagine you're standing in the shallows on a beach, counting the wave peaks as they wash by – maybe one every second or so. Now instead of water waves, imagine you're counting the peaks of electromagnetic waves as they wash by – so many per second of this type, a different number for that type.

If you were counting gamma rays washing by (assuming they didn't turn you into the Incredible Hulk), you'd count something like 10 billion billion per second. Fast counting! If you were counting X-ray waves, you'd count fewer: maybe only a billion billion. Fewer still ultraviolet waves would pass per second, perhaps only a few million billion. Now we're slowing down our counting into the region of visible light, starting with violet. From violet, you go backwards through the rainbow, right through to red and you'd count fewer waves per second for red light than blue.* Beyond red sits infrared or IR, the magical 'invisible' light that makes your TV remote work. Fewer IR waves arrive per second (about 10,000 billion) than visible red. Hopefully you get the gist. Beyond IR, with even fewer waves passing per second, is a range of electro-magnetic waves called . . . you guessed it! Radio.

Let's add in a useful word here: frequency. It's simply a measure of how many waves pass a particular point per second, and the 'per second' bit is also called hertz. So IR has a frequency around 10,000 billion hertz. Also UV light has a higher frequency than blue light, radio 'light' has a lower frequency than blue light, and so on.

Radio waves range from very roughly 300 billion

---

*If you ever need to remember the rainbow backwards for a pub quiz, try: 'Vain In Battle Gave York Of Richard', or perhaps 'Vaguely I Battled Giant Yaks On Roundabouts'.

waves per second (a frequency of 300 gigahertz if we're being scientific) to fewer than 10 waves per second, or 10 hertz. That's actually a huge range, with one end having a frequency billions of times higher than the other! For comparison, visible light, which means every colour you can see, covers only the range from just under 800,000 billion to 400,000 billion hertz. That's about a factor of two different!

All of these different electromagnetic waves form the electromagnetic 'spectrum', with ultra-high-frequency gamma rays at one end, and low-frequency radio waves at the other, and the colours of the visible rainbow somewhere in the middle.

Making waves

OK, here's another bit of physics. We're going to show how the number of waves per second (the frequency) is related to the length of the wave from one peak to the next (literally 'wavelength'– sometimes scientists run out of cool names).

Imagine you're standing on a motorway bridge watching a very weird traffic situation. Passing by underneath you, all at a constant speed, is a long, long line of identical giant trucks. They're all nose to tail because they really want to get to the next service station for a cuppa tea and a piddle. You stand there for a minute and count them: sixty trucks per minute, or one every second. Weird!

Now imagine that in the next lane, right next to the trucks, is an impossible line of Smart cars. They're keeping pace with the trucks perfectly, and they are nose to tail too, on the way to some strange Smart car festival. You count these as they pass by, and in one minute you get up to 600 Smarts! The frequency of the trucks travelling past you is 1 per second (1 hertz) and the frequency of the Smart cars is 10 per second (10 hertz). All the trucks and Smarts are travelling at the same speed; there must be a cop with a radar gun up ahead.

The only difference between the numbers of each vehicle you count in one second is how *long* each type of vehicle is. A single truck is long, but you can squeeze ten Smarts to fit in the same length of road.

What we're talking about here is exactly like light and radio wavelengths, except trucks, alas, don't travel at the speed of light.

Look at it like this. The trucks in our story have low frequency because they are long. The Smarts have a higher frequency because they are short. In exactly the same way, radio is a low-frequency electromagnetic wave because it has a long wavelength, like the trucks. Ultraviolet light is a high-frequency wave, because it has a short wavelength. It's the Smart of the story.

So how big are these 'wavelengths' for electromagnetic waves? Green light, which your eyes can see, has a wavelength of around 500 billionths of a metre, 2 million times smaller than a Smart car. Freaking tiny! Your home Wi-Fi emits and receives radio waves that are probably around 10 centimetres long. But that's nothing: traditional radio, er, *radio* waves can have wavelengths of over a kilometre.

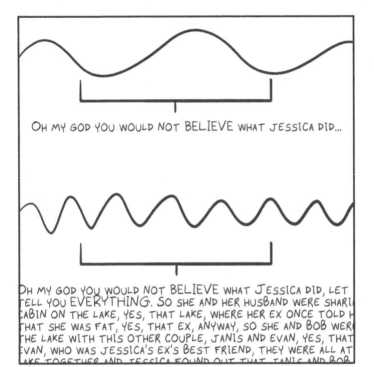

OH MY GOD YOU WOULD NOT BELIEVE WHAT JESSICA DID...

OH MY GOD YOU WOULD NOT BELIEVE WHAT JESSICA DID, LET TELL YOU EVERYTHING. SO SHE AND HER HUSBAND WERE SHARI CABIN ON THE LAKE, YES, THAT LAKE, WHERE HER EX ONCE TOLD H THAT SHE WAS FAT, YES, THAT EX, ANYWAY, SO SHE AND BOB WER THE LAKE WITH THIS OTHER COUPLE, JANIS AND EVAN, YES, THAT EVAN, WHO WAS JESSICA'S EX'S BEST FRIEND, THEY WERE ALL AT AKE TOGETHER AND JESSICA FOUND OUT THAT JANIS AND BOB

THE SHORTER THE WAVELENGTHS, THE MORE INFORMATION
CAN BE SENT OVER RADIO.

Bands on the radio

Right, now we've got that science sorted, let's talk about sending data over the radio. Let's start with, say, your voice. Imagine you had a radio transmitter that was sending out waves at a single frequency. Far away, your friend has a radio receiver set to the same frequency (the same radio 'channel'). You both turn the equipment on. You've got a machine that listens to your voice and changes your outgoing radio signal to match its sound. Your friend has a different machine that spots the changes in the radio signal, turns them into electrical signals and sends them to a little loudspeaker. Bingo, you've invented the walkie-talkie. Very cool.

But at that one radio frequency, there's only enough 'space' to send your own voice. If someone else sets up a similar system and chooses a frequency the same as yours, you can't both speak at the same time. If they use a frequency that is very close to yours, there will very likely be some overlap, thanks to a funky bit of physics that makes the radio waves spread out. You'll both get interference. For both of your walkie-talkies to work, you need to have two different frequencies far enough apart that they don't interfere.

Your system has another problem – it can only send your voice in real time. Let's solve that by adding a computer at your end that turns your voice into numbers (digitises it) before it gets sent over radio, and another at your friend's end of the link that can turn numbers into sounds. Now you've got a digital voice link, and if you're clever enough, you can use the computer to squeeze down the amount of numbers needed for your voice signal so it doesn't take up every moment of your transmit time. Now you can also squeeze other information into your signals, like snippets of text. Welcome to (a very basic form of) digital cell phone!

But what happens when you want to send much more data over your radio connection, more than you can fit into that one frequency you're using? The answer is to use a selection of frequencies right next to each other – each frequency different and far enough apart that they don't interfere with each other.* This is called a 'band' of radio frequencies, and it's obvious that you can use it to send much more information over your crude 2G radio link to your friend, as long as both of you have fast enough machinery to transmit and receive all those radio signals at once.

---

*Don't worry about why this happens exactly; just think of it being like stages at a music festival: if they're too close together, the tunes get all mixed up, so you keep them apart. Spacing out radio signals in frequency is a bit like this.

You should be able to guess now that the more radio frequencies you use concurrently – the bigger the band's width – the more information you can send at the same time. See? Bandwidth has a real scientific meaning; it's not about how wide Coldplay are if you line them up on stage. And it's far more useful than the annoying 'I don't have the bandwidth for that right now!' that overworked executives use as a terrible excuse.

Radio rainbows

Radio waves tend to be good at specific tasks depending on what their wavelengths and frequencies are. Basically, the very long radio waves with very low frequencies have some specialised uses compared to the very short waves with very high frequencies.

Some of the usefulness of different wavelengths depends on how they interact with other materials – for example, some radio waves bounce off the ionosphere, a weird electrified layer of our atmosphere far above the surface, so they're useful for very long-range communication. Other radio signals get scattered or absorbed by water molecules, making them ideal for weather radar systems – so that we can constantly see where clouds are forming and where it's raining and so on – but rubbish for air traffic control!

Another factor in radio waves of certain frequencies having certain uses is how easy it is to make an antenna that can transmit and receive them, since the size of the antenna is influenced by the length of the wave. Your Wi-Fi router's antenna is around 5 to 10 cm long, for example, but so-called 'long wave' radio masts can be a hundred metres tall.

Over the years, different sections of the radio spectrum have been given various 'official' names. The UHF band, 'Ultra High Frequency', used to be used for old-fashioned TV. UHF waves have a long range, and while they can get blocked by things like hills and large structures, they do go through walls effectively enough to make indoor TV antennas work. The exciting-sounding 'X-Band' (so named by NATO because . . . er, well just because) is used by modern military radar systems, and is also used to talk to spacecraft exploring other planets in our solar system because it can go right through our atmosphere. ELF radio (Extremely Low Frequency) can go around the curve of the earth and penetrate water, and has been used for atmospheric research and even to communicate with submarines.

The naming conventions are weird, and you can mostly ignore them. The important takeaway is that there are lots of reasons why certain radio frequencies are used. The other important thing to mention is that because of their different usefulness, lots of the bands available in the radio rainbow are already full.

*

Which leaves us where, exactly? you lament. Lucky for you, all is about to be revealed . . .

Chapter 4

So WTF Actually Makes 5G Different?

Now that you understand some of the radio science, let's talk about what's behind the technology that makes 5G actually work. By the end of this chapter, you'll be in a good place to see through some of the nonsense that's written about 5G, and maybe you'll begin to grasp why it's such a powerful innovation. There will be more of that in the next chapter too, which is a good one and has sex in it.

There's no one 5G to rule them all

The first thing to know about 5G – or any of the Gs for that matter – is that there is no 'one thing' that you can call 5G – it's not like the nose on your face, which you can point at and say, 'That's my nose.' 5G is a *collection* of planned systems that will bring about the next generation of mobile broadband technology.* In the nose analogy, 5G is the same as creating a list of what next-generation noses should do: have bigger nostrils, smell more smells, not get in the way of seeing . . . and so on. It's a list of functions, or criteria that have to be met, rather than a description of any one thing.

---

*Notice that I didn't say 'phone network', though I'll slip up for sure and use the term elsewhere in the book. 5G has *so much* Internet technology built into it that in reality it's a wireless network for high-speed mobile broadband that also happens to still allow phone calls and traditional SMSs, if those are your (slightly old-fashioned) cup of tea.

That list of goals for what 5G *should* be were drawn up by the Third Generation Partnership Project (3GPP). The 3GPP is a group of telecom standards bodies from around the world, and their goals include standardising things like data speed, radio frequencies, latency, connection density and so on. They're like the NATO of phones, making everyone play nicely together. Collectively, 3GPP's standards describe what's called 5G New Radio, or 5G NR. It may sound highly bureaucratic, but it seems to have worked out OK for 2G, 3G and 4G, which also were standards drawn up by 3GPP.

The point I'm making here is that there's no one single magic 5G ingredient that will be put in phones or phone networks to make them 5G; instead there are lots of ingredients.

Clearing out the radio spectrum

With every generation of mobile phone that's arrived so far, more and more people have become connected to the networks. With 1G cell phones there was barely enough bandwidth in the chosen radio wavelengths and it ran out. When more connections were needed, a leap to 2G digital had to happen because digital signals made more efficient use of the available bands. As later generations of cell phones needed still more bandwidth for more connections and faster speed, we had to make more room in the radio spectrum for these new services. So telecom organisations around the world began to clear out the radio spectrum of legacy signals, freeing up the space for mobile phones. It really was a case of saying things like: 'Almost no one is using terrestrial TV signals nowadays; let's turn them off!' and then putting up with some grumbles from people who still used the 'old' wavelengths for special purposes.

This habit has carried on over the years, with more users expecting faster network speeds, thus demanding more bandwidth, more radio space, for the cell phone networks.

5G NR takes this process to a whole new level. It says that there are going to be two main bands for 5G phones. One will be made up of a big chunk of the spectrum ranging from 0.4 gigahertz to 4.7 gigahertz. The other will use frequencies from around 25 gigahertz to over 50 gigahertz, a band that is about 25 gigahertz wide and new territory for phone network use. That's a lot of numbers to throw at you, but the thing to note is that basic 4G *only* used frequencies below 6 gigahertz.

To make space for 5G, countries have had to Marie Kondo the hell out of their existing radio licences and really chuck older stuff out. In the UK and US, for example, some digital freeview TV broadcasts have been moved to other frequencies to make space for early 5G network signals. Meanwhile, the Federal Communications Commission (FCC), which is in charge of the radio spectrum in the US, has told satellite TV companies that currently use C-band radio signals that they have to stop doing so by 2025 at the latest, so that this band is free for later use in more advanced 5G networks. Dramatically, this means that some satellite TV companies will have to launch totally new satellites that use different transmission frequencies in order to keep functioning. Actual rockets will have to be fired into space to make more, um, space! The Pentagon has even agreed to free up some of its radio frequencies, previously used for military communications, in order to make space for 5G and beyond.

Clearing 5G spectrum is a complex ongoing job, but governments want to do it, because by selling licences to 5G network operators they can make huge piles of cash. In August 2020, the US government earned $4.5

billion just for licensing a tiny slice of spectrum around 3.6 gigahertz. That's a *huge* pile of cash.

The upshot of all this preparation work is that 5G networks will operate across a much bigger range of the radio spectrum than earlier networks, but the specific regions used should end up being mostly similar around the world. The lower chunk of 5G spectrum has frequencies pretty much the same as current 4G networks, which is already making it easier for network providers to roll out 5G. Although the data speeds aren't expected to be very much faster than 4G, the latency is better and these radio signals have a greater range from the phone towers, and thus better connections, which is a big part of what we really want.

So-called 'middle-band' 5G is at the top end of the lower chunk of 5G spectrum, and it will use micro-wave* radio around 3 gigahertz frequencies and

*Argh! Microwaves! Like what I do to reheat last night's curry! No . . . don't panic. It's not that kind of microwave signal.

have faster speeds. But these radio waves lose their energy quite quickly as you get further away from the antenna, so the radio masts will have to be closer together. This kind of service will likely be the 5G found in some urban areas.

The most speedy, exciting bit of 5G, though? The bit that really takes it to the next level? That bit uses the 'high' band, based on a much newer type of radio tech that operates in that new territory I mentioned above. These are radio frequencies so high that the waves have a wavelength only milli-metres long.

Tiny waves, big news

If this was a certain kind of shock TV exposé, at this point the script would have the stage direc-tion: *cut to dramatic music and scenes of terrified punters fleeing screaming from the wildly swinging*

doors of a mobile phone shop, trampling on news-
papers that have headlines like 5G MILLIMETRE
WAVES CAUSE PANIC! *as they run down the street!*
Millimetre wave is the new bit, the big bad boy of
5G, the 'scary' bit that attracts tabloid headlines.
New things tend to scare people (and scaring
people makes tabloid newspapers lots of money).

Except, think about it for a moment and remember my physics lesson. Millimetre wave, mmW for
short, means radio waves that are millimetres in
length. On the electromagnetic spectrum, mmW
radio waves fit in between infrared light, which
measures up to a millimetre in wavelength, and
microwaves, which have centimetre-sized wavelengths. There is nothing terrifying about the *size*
of these radio waves by themselves.

OK, so the size is not important. But what do millimetre waves do? Plenty of technology already exists
that uses mmW, just not necessarily for

communications. You may even have used it your-self – it powers some of those giant walk-through full-body scanners at airports. There are two varia-tions of these machines. The first variety uses mmW transmitters and separate detectors to bounce mmWs off the skin of the person being scanned to reveal if they are carrying something illicit under their clothes – like a weapon, drugs or a chunk of rare unpasteurised cheese. But the other kind of mmW scanner doesn't have transmitters at all. It uses what are called 'passive' sources of millimetre wave radiation, coming from the environment or even the body of the person in the scanner.

Cut to scenes of people running and screaming again as they panic that their body is radioactive or something equally scientific-sounding and horrific.

You, yes, you sitting there. You really do – really, honestly, I'm not kidding – emit some forms of millimetre wave radio, though this shouldn't actually be much of a surprise. Some mmWs are, after all, quite close to infrared waves in wavelength, and our bodies definitely emit IR, because infrared cameras and those fancy 'no touch' medical thermometers that became so common during the coronavirus pandemic see the IR heat signals coming off our bodies.

So, mmWs are used in body scanners. They are also used for military purposes like radar. Radio astronomers have made some amazing discoveries about our universe using mmWs. And they are emerging as a scanning technology that could help self-driving cars 'see' using radar. Some telecom companies are even using highly specialised millimetre wave radio systems to help their existing mobile phone networks work, a parallel to laser signals shooting through fibre-optic cables and so on. Obviously not all these mmW uses involve the same frequencies, and only a bit of the millimetre band is going to be made available for 5G. The really neat thing, though, is that unlike lots of the rest of the radio spectrum, this band is relatively unused. So there's much less fuss involved in clearing out space for 5G than at other wavelengths. No Kondo-ing required; much joy is sparked!

Why are millimetre waves the next big thing for mobile phones? It all comes down to a bit of physics and a bit of maths that simplifies to this statement: the shorter the radio wavelength, the more information can be sent at once in a radio signal. More information per second means faster data arriving at or leaving your phone. MmW signals are millimetres in size, compared to the 50 cm length of 0.6 gigahertz radio waves (a frequency close to typical 4G signals). So, simply put, a mmW cell phone signal can carry more information at once than a 4G one.

MmW antennas can also be made smaller than is needed for other radio signals, so you can fit more antennas on a phone tower to let many more devices connect at the same time. Or you can deploy many smaller antennas around different parts of buildings like sports stadiums.

There are a couple of issues with mmW signals, though. Firstly, when I mentioned mmW scanners in airports, I pointed out that the waves 'bounce' off your body instead of passing through it like X-rays do (mainly because mmW signals are lower energy than X-rays, and your body is quite dense and watery). You can guess that similarly mmW phone signals won't pass easily through your body, which includes your hand when you're hold-ing a phone. They also don't like dense materials like bricks or other building materials, or metal reinforcements. But they can go through glass, and they do bounce around a lot once they are inside a room, which can actually be useful to spread the signal out. MmW signals can also suffer from 'rain fade', where very wet air absorbs and scatters some energy from the signals – GPS systems sometimes encounter this problem too.

If all of that sounds like a list of reasons why mmW technology is actually not very good, don't worry. There are plenty of relatively easy fixes for all of the downsides, like carefully designing multiple antennas into your future 5G phone so the signals aren't blocked by your sweaty palms. Plus I haven't even mentioned the best bit about mmW yet. So let's do it now.

Super-villain radio antennas

Perhaps the coolest, most science-fiction-sounding bit of 5G is 'beamforming'. This very nifty radio physics trick does exactly what it sounds like – it's about shaping the beam of radio waves from an antenna. To understand it, you have to let go of any ideas about a radio antenna being one of those traditional single metal pole designs, possibly with a little bobble on top (you know, the kind of thing kids draw poking out of the top of robots' heads),* or even one like a TV satellite dish. These traditionally shaped antennas can send or receive signals either from one direction only (the TV dish pointing at a distant satellite), or from all directions equally (like an old-fashioned radio antenna).

But a beamforming antenna is made of lots of smaller antennas connected in an array instead of just one antenna doing the job all by itself. I could

*You know the kind of drawing I mean. It's weird how children draw like this. I mean, they don't really see antennas like that nowadays, and yet there they are on the robot's head! It's like those houses with four windows and curly smoke from the chimney that kids also draw. Definitely a topic for a good episode of *The X-Files*.

try and come up with an explanation here using a cool story about the way certain villains' evil laser beams join together to make a more powerful beam shooting doomily off towards a planet in a famous sci-fi movie, but fun as this would be, it's not *quite* how antennas work. Instead, think of an antenna array working a bit like a super-fancy modern showerhead, with water shooting out of lots of tiny holes, versus trying to wash under a garden hose or perhaps a waterfall, which just blasts lots of water out willy-nilly.

By varying how the outgoing radio signal goes to each tiny antenna, you can cause a very tightly shaped radio beam to shoot off from the array in the direction you choose. Or you can cause many such beams to go radiating out from the array in different directions. In the showerhead version of how this works, it's a bit like having an even fancier super-efficient, eco-friendly 'massage' shower-head. By changing the control dial, you adjust how the water goes to certain holes in the shower-head, so the water either jets out in several directions, or very precisely in just that one direction you carefully choose. I have to point out here that this is very much *not* how radio antenna arrays actually work – they are fantastically complicated and have super-clever circuitry and run on hard maths done very fast – but it's a pretty damn good sketch of the idea.

Antenna arrays work the other way around too: by processing the incoming radio waves from the whole array, it can 'listen' in a very carefully chosen direction for incoming signals. The listening direction and strength can be changed from one second to the next, and all this happens without any moving parts (so it one-ups the fancy-schmancy showerhead there!).

It all sounds fantastically complex, but the technology has been used for decades in all sorts of radio systems, including military radar systems and even some advanced home Wi-Fi set-ups. But why would we want this for 5G? It's easiest to explain if you think about power. A 4G antenna has to send out radio waves in all directions to talk to a single phone because it doesn't know *where* that phone is. That uses up more power than if it could only send out radio signals exactly in the phone's direction. But a beamforming antenna can send radio signals precisely where they're needed, which can use less power overall. Less power means less electricity, which is great for the phone company, the environment and ultimately for you! Basically, the old antenna system uses lots of energy to cover an entire area; the new array uses much less energy by being very selective about where it sends signals. There are also a bunch of other benefits as well as potentially reducing the power needs of a radio mast. These include less chance of interference of radio signals between users, avoiding issues like signal fade caused by buildings in cities, and helping more users connect to the network per radio mast.

In real life, beamforming technology is a bit more complex than this, obviously (showerheads . . . ha!), and the term describes several similar techniques that involve processing radio signals for lots of tiny antennas at once.

Oh, and one neat thing. Beamforming works particularly well for mmW radio signals at consumer scale, because you need smaller antennas to make it work, and can pack in lots of these as an array in a box on a phone tower, taking up much less space than if you tried to do the same for 4G signals.

The small size and relatively low power demands of mmW antenna set-ups means we may end up seeing beamforming antenna arrays everywhere. They should be quite easy to install in shopping malls, on individual floors of tall buildings, at strategic points around sports stadiums or other large venues and so on. So while the 'low' band of 5G will cover large areas using similar infrastructure to 4G, it seems like the speediest part of the technology could become ubiquitous in urban areas.

*

The goals for 5G NR don't stop with more spectrum, millimetre waves and tech like beamforming, of course. They include specifications about speed, multiple connections and latency that I mentioned in an earlier chapter. 5G NR also includes standards for different ways to process and encode the radio signals, better security protocols and much, much more, with the overall goal of being a vast improvement on 4G.

And now we're going to dive into the impact that all this cool science and technology will have on the world.

Chapter 5

5G Up, 5G Down, 5G Twirling All Around

There's been a lot of very excited prose written about 5G, suggesting it's so amazing it might change everything. But . . . that's a bit vague, and the question remains: what will it actually do for you when you get it? Time to hold on to your hats again, because we're off on a speedy tour of the positive things 5G may bring.

I'll also take a quick look at the alleged negative aspects of the 5G revolution, since you may feel like many other people and worry about 5G's impact on your life. It's important to consider this, since 5G is going to impact billions *of people's lives. Including writers who have got tangled up writing an intro paragraph to appease their editor . . . That's me!*

I just called to say Five Gee

OK, if I've done my job properly, you'll know that the best thing about 5G is going to be reliably high data speeds. The first 5G tech you will probably hold in your hand will be a 5G smartphone that is pretty much like your old 4G one, just faster for downloads. But as the roll-out of 5G continues, the phones will get even faster. And faster. Faster downloads, faster uploads, faster everything! Picture this:

```
You're in a city of the near
future. You jump out of the train
```

and into an Uber to get to an important meeting. As the car threads through traffic, your phone rings – it's the six-way video conference you're supposed to be attending. It's moved an hour earlier! In 2020, that could be hugely stressful. With 5G's super speeds? Not a problem: video clear as a bell, no lag, and solid connection even among all the tall buildings. You may have played an intense video game on your 5G tablet on the train, though, so you might not be as prepared for the meeting as you should be . . . but, hey – 5G has its downsides!

I've used the word 'phone' a lot in this book. But 5G really isn't about telephony, and Alexander Graham Bell is probably spinning in his grave at this (ahoy there, Alex!). 5G is really enhanced mobile broadband. It's about data, all the time. That means it'll be found in many more things than mere phones, carrying on the trend started by 3G. 5G will be in your tablet computers, your laptops, hybrids, USB dongles and so on. It may even overtake the use of Wi-Fi in some cases, where the network speeds and reliabilities are fast enough or the location otherwise suits it. Most everything that is computer-like that is connected to 4G will switch to 5G over time, and be faster for it. And thanks to different 5G technologies like very low power connections and smaller antennas, you may see 5G commonly used in places that 4G is rarely used today – like in smartwatches. And that's just the beginning . . .

If you like it you'd better put the Net on it

Another tech term that's thrown around in the media is IoT, or the 'Internet of Things' as I mentioned before. It's jargon, but the idea behind IoT is that at some point a connection to the Internet will be included in pretty much *everything*: even the proverbial kitchen sink could become 'smart'. Imagine the possibilities!

```
You arrive at your front door tired
after a frantic day at work. The
lights are already turning on, the
door magically unlocks as you get
close - because your phone has been
chatting to all the smart devices in
your home, and when you turned on to
your street and walked through your
gate, everything IoT inside knew you
were coming home. Your favourite
```

relaxing tune may be playing as you slump on to the kitchen stool, the lights in the room may fade to match the mood of your most recent social media discussions, and your smart coffee-maker may turn on to deliver an espresso – or, depending on your preference, your smart fridge may serve up a freshly poured glass of rosé. OK, this last bit is a fantasy, and the whole idea sounds like a slick ad for a high-tech start-up with a name you can barely pronounce. I'm also sure it'll be years before your 5G robot dog brings you your (smart?) slippers as you sit on the (smart?) sofa, but lots of the rest of this home-based IoT tech is already here or on its way soon. You really can buy smart coffee machines!

Almost everything that is currently a smart IoT device relies on one of two main technologies to connect to the Net, which is the magical source of the 'smarts': Bluetooth or Wi-Fi. Wi-Fi is an established technology and it's a relatively cheap and easy choice for IoT manufacturers. Bluetooth offers a similar benefit, but a Bluetooth smart thing usually needs a gateway like a phone or home hub to connect to the Net properly.

5G could totally change all of this, bringing IoT to maturity and replacing Wi-Fi and Bluetooth in tons of new IoT devices. The people who designed 5G claim that it should allow 'massive machine-to-machine connections', which means 5G is designed from the start to make IoT better. 5G's lower frequency band is an excellent choice for this because the signals can go a long way, and while this band doesn't have speeds like mmW 5G, it is pretty much perfect for a whole new class of IoT things that are either mobile or

REJECTED IOT DEVICES

permanently outdoors. Exactly what these things will be is anyone's guess.* But basically, if you can imagine a device having a battery and a Net connection, it could soon become a 5G smart thing. Everything from pet collars to squirrel feeders to smart watering sensors in your garden . . . to more powerful 5G 'key finder' tags.

The future's made of virtual reality

Virtual reality has been 'the next big thing' for, oh, fifty years or so. That's a pretty long time to be the next big thing without ever quite, in fact, becoming a big thing. Sometimes, for bleak years at a stretch, it has almost seemed like it won't actually

---

* If you can successfully guess what these next-gen IoT things are, right now, there's a billion-dollar market out there just waiting for you. No kidding. It's perfectly possible that the next Apple or Tesla is just around the corner, with some crazy IoT 5G device that no one has even imagined yet.

become a thing at all. The reasons why could probably take up a whole book of their own, but they boil down to the fact that for a long time, VR technology wasn't able to live up to the dream.

But VR is a stunning idea. It promises to immerse you in a computer-generated 3D world that feels much more real than watching a flat screen. Recent advances in screen tech, mobile processors and motion sensors (similar to the tech inside a good smartphone) have begun to bring VR to the masses. But it is still not a mainstream thing, possibly because for some uses you need a powerful computer to drive the VR goggles. Mobile-only technology isn't quite up to the job, and some wireless data connections are too laggy. Lag is annoying in regular video games, but for VR gamers lag and slow video can lead to all sorts of fun side effects, including motion sickness!

*Enter 5G, like an enthusiastic single at a speed
  dating event*

5G: Hello!

VR: Er, hi . . .

VR narrows its eyes as it sips a cosmo

5G: (*Creepily*) I'm your future!

VR: You are? (*Sceptically, after forty years of
  loneliness*)

5G: Yes (*Said as 5G turns on its speed, low latency
  and massive connectivity powers*)

VR: Oh my!

5G has so much promise for improving VR tech
that some chose to rename it 'XR', which seems to
stand for 'extended reality' or 'cross reality'. XR
covers both VR, where reality is replaced by
graphics, and AR, or augmented reality, in which
your view of the world is 'enhanced' with extra
information and graphics. It also covers mixed
reality, where virtual objects and worlds interact
with real-world information in real time. XR truly

is the kind of thing science-fiction writers have been banging on about for decades. 5G's high speed and low latency is ideal for this, with little to no delay between the headset wearer moving their head and the digital view of the world being changed. You could do this with ultra-fast Wi-Fi, but here's the *other* transformational thing 5G adds to XR: XR headsets will be able to go outside into the world. No longer will gamers be relegated to dark basements. Though come to think of it, whether they will actually *want* to leave their basements is another thing entirely.

What techno marvels XR will enable are still unknown. We can imagine some, though . . .

```
Once you've sipped your smart-
device-delivered coffee or downed
the nicely chilled rosé, you may
choose to de-stress from work with
some virtual tai chi. You step into
```

your back yard, breathe the fresh evening air, and on goes your XR headset. With a word of command, the gritty urban real-world view is replaced with a soothing mountainside, and gentle wind sounds hit your ears. In the background, cowbells. The headset measures your body position, spots where your hands and feet are moving, and guides you through a relaxing half-hour session, only briefly spoiled halfway through when a video call from your mum pops up on the display. It's OK, you can speak to her later.

VR gaming is already booming within the limits of current technology, thanks to companies like Facebook and Sony. Meanwhile, AR tech from companies like Microsoft is being used in all sorts of industries, including in advanced car design and even manufacturing. On a more consumer-facing level, Apple, with billions of iPhones and iPads around the world, has built limited AR technology into its operating system. Companies like IKEA have made use of this, so now it's possible to use AR to test out your choice of sofa right there in your living room without having to visit a store.*

Will 5G bring us Hollywood-level VR or AR movie experiences? Will education embrace AR to make science and history lessons stunningly fascinating, with immersive graphics depicting dinosaurs chomping on each other, or stars being torn apart by black holes? We have a few answers to that

---

*Technophile readers may be waving their hands frantically right now, trying to point out that Google launched a VR headset called Glass that spectacularly failed to change the world. It's possible, though, that people just weren't ready for this sort of innovation back then.

THINK OF ALL THE THINGS VR CAN HELP YOU EXPERIENCE...

already: during the coronavirus pandemic, Kansas City University in the US made headlines by using VR to help train medical students, and Coventry University in the UK held virtual seminars using VR transmitted over its own low-lag 5G network. School may never be the same again!

Can you hear the people drone?

Wvvvzzzzzzzzzzzzzzzzzzz zzzz zzzz . . . When you hear that characteristic whiny buzz of a drone, chances are if you look around you'll spot its pilot somewhere nearby, wielding a bulky remote control. A lot of commercial and consumer drones tend to be more or less line-of-sight vehicles, with the pilot needing a pretty good view of what the thing is doing and real-time control over how it's flying. Otherwise they risk crashing it, or getting it stuck in a tree. Some drones, of course, have real-time video feeds to the controller so the pilot can see what the drone sees, and these can stray a little further afield, as can drones that are programmed to fly on autopilot. But on the whole, drones are limited by the radio range of the controller.

But all of that could change with 5G. Drones can send a ton of data in real time back to the controller, and over 5G that could easily include live high-resolution video. The reliable connection and low latency of 5G could even allow drones to rove much further away while staying under full control.

Drones have all sorts of uses, from security to high-risk camera work (like inspecting electrical power lines) to aerial photography for fun or surveillance. That last bit is where drones get controversial, because you feel very differently about drone surveillance if you're the person being spied on. Your reaction may also vary depending on whether it's a police drone, a news drone or a nosy neighbour's drone.

All sorts of drone uses could be improved by 5G, and possibly a whole new batch of uses could evolve. After all, who could've imagined ten years ago that high-speed drone racing would emerge as a thrilling and amazing new sport? (Go check it out now on YouTube – it's bonkers!) Today drone racing tends to happen in relatively small venues because of the need for swift control data and video feeds between the drone and pilot. With 5G, what if giant drone racing could rival Formula 1? 5G could even make all those drone delivery services a reality: in Spain, network provider Vodafone has tested a 5G drone capable of delivering emergency medical equipment like defibrillators really fast over ranges up to 10 kilometres – likely beating ambulances on regular roads.

Remember that video call from your mum? She called back, and it's a good thing you were outside already so you could wave at the friendly UPS drone and collect the package she was having couriered over. Thanks to 5G, it navigated precisely to your home, avoiding other drones, and the package was summarily dropped on to your picnic table. An early birthday gift from your mum. Sweet!

I'm not driving my car, it's driving me

That 5G is going to be included in cars of the future seems pretty inevitable. Our cars are already becoming semi-computerised machines with built-in tablets and entertainment systems, particularly if the car is electric-powered. Cars with basic Net connections have actually been around a while – whether to power your swanky entertainment and navigation system, update you on your vehicle's charging status while you're shopping, or automatically call the emergency services if you've had a crash.

5G in your regular family car will super-charge all of the 4G and 3G connectivity stuff that some cars have today. In-car entertainment will get cleverer, navigation will too, and your car may even offer things like in-vehicle Wi-Fi or perhaps a built-in transport-friendly version of Amazon's Alexa or Apple's Siri.

The really clever bits of 5G vehicle technology, though, may not be something you get to see as a driver or passenger. While it's all very neat that your car can connect to the Net so your kids can watch a movie for the five hundredth time, that's merely a form of car-to-Internet connection. 5G is designed to allow new types of connection, including vehicle-to-vehicle and vehicle-to-infrastructure (as in smart road signs or in-road sensors). Forget any *1984*-like dystopian fears, though, because this could radically change road use, and could be critical for automatically preventing accidents, or at least reducing their severity.

All of these benefits become even more obvious when you're talking about self-driving cars. Right now in self-driving tech, either rolled out in a limited form by Tesla or in research projects around the world, the car itself has to do some very intense calculations in real time. But a highly connected 5G car wouldn't have to do all the work reading speed-limit or warning signs, or planning ahead if it sees slow traffic. Instead, it could get that data in real time from the road system itself, or from reports directly from nearby cars.

You decide to visit the olds back
home to thank them for the birthday
gift, and you summon a self-driver.
It's only a 45-minute journey, so
you can binge-watch a couple of
episodes of your favourite show on
your tablet on the way. Ten minutes
into the ride, the car suddenly
slows and pulls off the road.
You've not been paying attention
and are about to ask it what's
happened, but it says that a truck
up ahead has declared an emergency,
causing a small traffic jam. The
road systems suggest the road may
be closed for a while, so the car
has decided on a new route for you
– unfortunately another 20 minutes'
journey time. It apologises. Sweet!
Time for one more episode.

5G cars, trucks, buses and so on could even be joined by more and more automated delivery vehicles, each using 5G to navigate and stay connected. What I'm talking about here is more than mere cars, of course. 5G could also bring us a totally new sight on the roads: the self-driving robot.

Old MacDonald had a smart farm . . . or factory

5G has got so many technologists excited that they're pinning the 'smart' prefix on pretty much everything you can think of. Smart cities could manage flows of traffic and pedestrians more easily and lower electricity use, and smart bins could alert the city to come and empty them.

Then there's the matter of smart factories. Imagine a vast factory, making the very best in widgets. Robot welders whirl, sparking like good

TV special-effects, as they put widget bits together. A conveyor activates, taking the widgets to the next robot, which glues or solders on other widgety bits. At the end of the flashing, buzzing and probably very clean and tidy production line, robot porters pick up the finished widgets, tuck them in boxes and ferry them off for delivery. Every bit of the robotic factory would be talking to the control room, and other nearby robots, over a local 5G network – including sending hundreds of high-res video feeds to the office, where just one human supervisor oversees it all. 5G's speeds and low latency really could help this come true, and if you've seen video of Tesla's robotic car factories, you can see the beginnings of this sort of system taking shape, albeit without 5G's benefits.

Now take these ideas and apply them to a farm – an al fresco food factory – spread over a huge area too big for a Wi-Fi network to cover. Crops in distant fields could be monitored by 5G cameras, and the soil they're growing in could have its chemistry constantly checked by 5G sensors. Watering systems would be automatically turned on via the network, or covers could be robotically pulled over delicate crops in bad weather. Some crops could even be harvested by 5G-connected robots, with the farmer sitting comfortably far away from all the fast-moving, dangerous machinery.

The idea seems strange, since when we think of farms we tend to think of idyllic fields full of green things waving in the wind, with perhaps a lone worker driving a tractor or a cow grazing. But a blanket of 5G coverage, even in rural areas, could really transform farming, and not just for giant county-sized operations – traditional farms could benefit from sensors reporting on crop readiness or even tracking wayward sheep and cows. It's a trend that's begun already with 4G, and it likely won't make a big impact on farming until later 5G innovations arrive, but with the huge amount of money at stake, as well as the tricky problem of feeding us billions of humans, it's certainly going to happen.

As your car wends its way through the ever more rural landscape near your parents' place out in the country, the latest episode of the show you're watching finishes, and you glance out at the fields rolling smoothly by and remember the words of that old nursery rhyme:

> Old MacDonald had a smart farm,
> One zero, one zero, one.
> And on that farm he had a smart tractor,
> One zero, one zero, one.
> Self-driving here, self-driving there
> It harvests autonomously everywhere . . .

Doctor, Doctor, gimme 5G

Rural communities around the planet, whether in more or less developed nations, have long known the advantages of basic telemedicine, whether it's over the radio or nowadays via broadband Internet. But the whole world got a dose of how important it could be when the coronavirus pandemic hit in 2020. Suddenly it made magnificent sense to be able to talk to a doctor in real time without being in the same building as lots of coughing sick people.

With 5G speeds, reliability and low latency, telemedicine will get much cleverer. It will be possible for large medical imaging files to zip between remote doctors, scanning equipment and patients as needed, even in very distant locations. Video conferencing with high-resolution real-time imaging means medical consultations could happen in places they never have before.

Long-term wearable health sensors could let medical technicians and patients keep an eye on the progression of diseases or healing of injuries and so on. Wearable medical sensors are going to become more common because the potential benefits are so evident: the data from these sensors will benefit more than just the wearer; it will help develop treatments for many other people too.

Before you scoff at 5G and healthcare, remember that the Apple Watch is already credited with saving many lives by spotting early signs of heart disease, and even calling emergency services when a wearer falls and is hurt. This is all with very basic health sensors and 4G connectivity – how this tech will advance with 5G is almost unguessable.

With 5G connections it's even possible that remote surgery, with the aid of robots and low

WHAT CAN 5G DO FOR YOU?

DOWNLOAD MOVIES IN MINUTES!

GET CRYSTAL CLEAR CALLS!

MOO?
MILK YOUR COW!

CLEANSE YOUR PORES!

Scrape Scrape
SPACKLE HOLES IN THE WALL!

BREAK UP WITH YOUR PARTNER SO YOU DON'T HAVE TO!

latency, could happen at the hands of the very best surgeon possible no matter where in the world the patient or doctor is at the time. 5G could literally get under your skin (with the remote guiding hand of an expert medic, that is)!

You're not far from your parents' place now, so it's a surprise when your tablet lights up with a call from your dad. Not a problem, he says, but he's just been on a tele-Doc call with that nice Dr McCoy – remember her from when you broke your arm? – and she says that they need to tweak his meds a bit. The pharmacy has already been sent an email – would you mind stopping at the village to pick them up? They could easily get a drone to send them over, but since you're passing through anyway, it's no skin off your nose; you just have to ask the car to go there. It takes a matter of minutes when you arrive for the pharmacy's 5G-connected ID systems to verify that you've been given permission to pick up the medications, then for the pharmacist to print out a VR label explaining the dose, and you're soon back in the car.

Sexy, sexier sex!

Sex! It's fun to say it, fun to do it, it's how the human race survives, so let's not pretend that 5G isn't going to have an impact on the sex lives of millions of people around the world. Let's face it, some of the biggest tech advances in recent decades have driven changes in our sex lives, and the same the other way around – sex tech has also contributed to some big billion-dollar business decisions over the years.

If you were in any doubt about the penetration of high tech into the deepest, most intimate corners of modern life, the COVID-19 pandemic should have fixed that. With large-scale social isolation, people who found themselves physically apart from loved ones, potential dates or even casual social encounters of all sorts turned to technology like Zoom, Skype and so on and even to traditional sexting to achieve some form of emotional and sexual closeness with another person. This wouldn't have been possible in any other era, and it was powered by home broadband and 4G.

Exactly what 5G will do to change sex is a pretty hard matter. But there are already sensational news headlines like 5G SEX DOLLS WILL BE HUMAN-LIKE! and I can envisage how the high data speed plus access *anywhere* that 5G brings, along with improved AI, will lead to more tech being found in our bedrooms, bathrooms, living rooms and anywhere else sex happens.

With an explosion in Internet of Things tech and robotics, 5G may even begin to change people's sense of what love and sex mean. We really are about to enter into a world where phrases like 'love and sex with robots' and 'teledildonics' are going to be less sneered at and, most likely, welcomed with open, er . . . arms.

And let's not forget the sticky issue of pornography! It's been a big part of the technology business since the invention of the camera, and it'll certainly get a 5G boost. Exactly what this will entail, no one knows, but the industry is an early tech adopter, and will firmly embrace what 5G has to offer. Maybe that XR experience will get a couple of extra X's?

After a mild day of bumbling around with your parents, eating a few too many cookies and drinking one too many glasses of wine ('It's good for the heart!' says Dad), you're heading home. It's date night with your partner, but they're overseas on business. You have an idea, and make a quick call from the car. It might be thanks to the wine, but you feel like shaking things up a bit and coyly murmur a suggestion. The moment you get home, you call again, slip on the 5G XR headset and . . .

No, I'm not going to tell you what happens next.

Itsy-bitsy teeny-weeny 5G radio cells

One bit of mobile phone technology you may not have heard of is 'picocells'. These are basically small cell phone towers that can be installed in certain places to boost the signal of an overall network or increase coverage in a very precise area, such as inside a shopping mall or even a passenger jet. Their coverage area is usually only a few hundred metres, but the equipment is small and simpler to install than a full antenna tower. There are smaller-range versions called femtocells, with a range of a few tens of metres, and larger ones called microcells, that can cover up to a couple of kilometres. In some countries, the installation of these small cells may not even need a government radio licence, because their range is so limited and, particularly for mmW, the spectrum is relatively uncluttered.

5G picocells and femtocells could, thanks to mmW 5G, be even easier to install than before because the antennas are so much smaller. At a trade show years ago, Ericsson even had a roll-up prototype antenna array that you could just drop on the floor and that was pretty much the installation!

On your journey into work the next day, you choose the underground for a change. Thank heavens you did, because as you're buffeted around in the carriage, your smartwatch beeps to say it's just started raining. A lot. Glad that you've got a good network connection down here, you call ahead to your colleague and suggest you move the morning meeting to the coffee shop instead of the office. That way you won't have to dash across the

```
courtyard in the pouring rain (but
he might, LOL).
```

After all these bits of narrative, I bet you're thinking I've read too many breathless bits of public
relations material put out by the 5G industry. Let
me tell you, my inbox really is full of them! But to
show that I know there is a downside to 5G, let's
look at it not as a beneficial tech, but as one that
could have undesirable side effects. Note I didn't
say 'will have', though, because alongside the
sharp criticisms of 5G that do have some merit,
there are also some wild conspiracies!

The digital divide

Our human world is horrifyingly unequal.
Because sometimes humans, I'm afraid to say,
even with our shiny amazing tech, suck. Regions
within countries, urban versus rural areas, whole

cities, nations or continents: the digital divide between those with easy access to technology and the Net and those without can be very sharp. As our world becomes more and more digital, this divide becomes sharper and cuts into some people's lives deeply, affecting access to health-care, education, banking, government services and more.

So far this book has been written from a developed world point of view, and we've not mentioned the social angle of 5G. So here it is, all in one big emotive whack, no joking, no punning:

We simply don't know whether the rollout of 5G will widen the digital divide or help narrow it.

One theory is that 5G will widen the divide by giving technologically advanced areas faster and more reliable connections, bringing with it the economic benefits of technology revolutions. On the other hand, in places with little existing digital infrastructure, installing 5G could be easier than in the crowded streets of, say, New York or Paris, and the arrival of even basic 5G systems could transform the lives of millions of people much more dramatically than you or I merely being able to watch mobile Netflix in buttery smooth 8K.

No one likes having small print

Sometimes the way phone network operators carry out their business can feel like those dodgy games of 'Chase the Lady' played by old-style street swindlers. Like when you expect unlimited Internet on your phone, only to discover (after downloading a whole series of a Netflix show) that what you really have is quote-unquote-unlimited and that the buggers have included some small print that says *Note: Unlimited doesn't mean without limits and does, in actual and legal fact, mean really rather limited. And when you go over the limit we'll say, 'All right then, hand over the cash. Ner ner na ner ner.'*

There's no guarantee that phone networks won't play some unfair or misleading games and shenanigans with 5G services that end up costing users lots of their hard-earned cash. When 4G arrived, there was a good deal of fuss about this sort of thing, because *technically* to deliver 4G cost the networks less than 3G, but weirdly, can't imagine why, the savings often weren't passed on to the paying public.

When there's a big paradigm shift like this, you have to watch for people making money out of it. It's just in the nature of business, sadly. In August 2020, for example, US network Verizon made headlines and then backed off plans to charge $10 extra for 5G access. Expect to see more of these shenanigans.

Radio interference

Remember that one big difference between 5G and earlier mobile phone systems is exactly how much radio spectrum is being given to 5G – its bands cover many more frequencies than 4G's did. There are legitimate concerns that some 5G bands are very close to frequencies already in use for other reasons, such as radar bands used for weather spotting. Suddenly having lots of noisy 5G phone signals in this radio region could upset weather forecasting! If you follow the logic of some critics, this could then impact the economic stability of many industries and even affect national security!

That's some pretty aggressive drum-banging based on not a lot of practical experience yet – the roll-out of 5G is only just beginning, and there are protections in place to prevent interference. 5G spectrum usage will inevitably result in interference issues in some places, but we just don't know how much of a problem it will be. Luckily these problems aren't unsolvable, as 5G phones and networks should be able to use a different frequency.

E-waste and the environment

The switch to 5G has environmental consequences too, though whether they will be a good or bad thing overall is unclear. Any big generational change causes people to switch to new hardware, which means potentially millions of tons of new e-waste full of chemicals and hard-to-recycle parts as consumers ditch their old 4G phones. Equally, if phone networks have to make big infrastructure changes, that could also result in plenty of industrial e-waste.

But the strange and wonderful thing about 5G is that while we describe it as more 'powerful' than

4G, it may end up needing less actual power to run. Some academic studies have suggested that 'idle' 5G networks may burn nine times less power than 4G LTE networks do. Pretty neat! The general idea is that 5G networks and devices should use less electricity than 4G. This comes partly from the power savings offered by beamforming and partly from the different way that the 5G radio masts keep in touch with phones in their range, which takes far less electrical energy than before.

There may even be energy savings, and thus a positive environmental impact, when 5G Internet of Things devices get embedded in infrastructure – a town in Portugal has installed some basic smart bins, for example, which are only visited by trucks when they're full, saving fuel, time and money!

Visual markers of 5G

Some people dislike the look of phone network infrastructure already, and there are already efforts made to hide it in some places by disguising phone towers as artificial trees (generally not very convincing, though!). And while we're kind of used to a world full of phone antennas, 5G networks will need many more of them. On the other hand, 5G equipment, particularly if you're talking about mmW, may actually be smaller than 4G and 3G versions, so it may be less visible. For this one, we'll probably have to wait and see.

Security

A big feature of cell phones is that each generation brings better security than previous

THE DANGERS OF A <u>TOO</u> WELL HIDDEN 5G ANTENNA

ones. It's not as sexy* an aspect of 5G as its super high speed and so on, but it is a critical bit of the puzzle. 5G has been designed with better security for the end user and in terms of protecting the infrastructure – all those servers, radio signal processors, and mysterious humming boxes in those little cabins near phone antennas.

The thing is, because 5G's got this massive connectivity thing, it involves much more infrastructure and will connect many more phones, including millions more Internet of Things devices. So there's going to be 'more 5G' for potential bad actors (as in bad guys, not dodgy graduates from backwater theatre schools) to attack. In security speak, the 'attack surface' for 5G could easily prove bigger than 4G has, which means there are simply more ways for the bad guys to get into 5G

---

*Sexy from an engineering, boffin, science-type perspective, at least.

systems. For example, many Wi-Fi or Bluetooth IoT devices are already notoriously insecure, and it's pretty typical for home users to leave default settings untouched – including passwords. You may think there's only so much havoc you could wreak by hacking someone's robot vacuum cleaner (though anyone who's had a Roomba smudge dog poo all over their floor may have pungent views on this), but it's more about how a bad guy could get into your private user details, including bank accounts and so on, *via* your Roomba. With 5G Internet of Things expected pretty much everywhere, from your key fob to the robots in automated factories, this sort of hacking may be a much bigger problem.

Lastly, let's not forget that security itself could be a security issue. There have already been one or two frightening stories in the news about people's smart home camera systems being attacked by hackers, including ransoms being demanded after smart front door cameras were hacked. All of this is just going to get much more complicated when camera systems are more ubiquitous and 5G-connected.

Spying, trolls and the unknown

One massive upside of 5G is, as I've mentioned a few times, that it's so powerful we don't know what amazing benefits it will bring into being! But I have to also point out that 5G is so powerful we don't know what undesirable side effects it will cause. It's a light side/dark side *Star Wars* situation really.

One giant downside, if you're a 'big government' sceptic, is that massive adoption of 5G technology like ultra-precise positioning or even private data held on apps could tempt governments to demand more access to citizens' 5G phone/Internet of

Things/smart camera data for reasons such as 'the public good'. In late 2020, a scandal hit the news when it emerged that the US military, among others, was purchasing totally legal location logs made by ad tracking systems inside popular apps, and tracking people all over the world! With 5G bringing expanded use of phones and IoT devices, this sort of problem is going to happen over and over again.

Another potential downside: home broadband has enabled some very dark behaviour by people with perhaps not the best morals,* sometimes protected by the anonymity of being on the Net. So massive 5G-scale mobile broadband could just make life easier for trolls and other online bullies. 5G everywhere means 5G connections everywhere, which

---

* Some people just aren't good, and deserve a description like 'Oh villain, villain, smiling, damned villain!', as Shakespeare eloquently put it. Or 'a wicked bugger' as my gran would've said.

also means trolls, hecklers, hackers, spies and worse everywhere.

Argh!

*

I've whizzed really fast through lots here – looking at some of the amazing benefits 5G could bring to the world, and some of the potential bad things that could happen too. Don't panic if you didn't understand some of them, or if they sounded scary. Gird your loins: there are actual 5G conspiracy theories that I need to tackle!

Chapter 6

5G: Sense and Sensibilities

The time has come, my friend, to talk of many
things. Of 5G and conspiracies and spying games,
signals and radiation pings.

*

We've come to the most difficult, and perhaps the
most important, bit of this book. It's a bit that's all
about questions. These questions about the really
ominous downsides of 5G, some armchair critics
say, are very important questions that are as yet
unanswered. Critics of the critics might point out

that it's quite hard to answer this sort of highly technical question if you're googling your research from the comfort of your armchair, but still . . . questions remain. These are the sort of big, scary questions that upset some people's sensibilities and lead to big, scary news headlines.

Is it secret?

One of the ongoing scandals surrounding 5G definitely has something of the James Bond, or perhaps *Mission: Impossible* about it. It concerns a young Chinese tech company founded in 1987 that in its relatively short life has achieved revenues in the hundreds of billion dollars, and whose technology has been intertwined through the networking systems of the world both behind the scenes and on centre stage since, oh, about the 2G cell phone era. You may have heard of this little outfit: Huawei. Let's go with the spy novel angle and call this whole issue of 5G secrecy 'the Huawei Factor'.

In recent years, Huawei has found itself embroiled in a conspiracy theory-like controversy that has reached far above gossipy newspaper headlines to the lofty heights of governments around the world. The details of why the company is suddenly controversial are complex and really thorny, but we can boil it down to this: for years, Huawei has made equipment that powers cell phone networks around the world, and naturally it was expecting to make the most of its expertise when it came to 5G. Then there arose a government-level worry that the company seems to have, ahem, cough, wink, uncomfortably close ties with the Chinese government. And since we're going to all rely on 5G to do so much more than we use 4G networks for, perhaps we should check that the Chinese government can't use Huawei's tech to monitor, influence or downright spy on whoever else in the world they want to. Before you know it, Huawei was the subject of bans, vetoes and regulation in many countries.

Has there been genuine intelligence work to discover if Huawei's 5G technology really does represent a threat? We don't know. And we might never know, because that sort of secret stuff is, well, secret. I suppose that if an ageing politician leaks something in a memoir in a few decades, or a whistle-blower's leaks lead to a Netflix documentary, we might ultimately find out the truth. Possibly. Huawei has vigorously protested its innocence, but the bans are in place and now other, presumably more 'trustworthy' companies are well placed to install much of the 5G infrastructure in big countries like the US.

Is it safe?

Oh boy. This is the motherlode of conspiracy-based news here. According to armchair critics, and other much more vocal critics who've even gone to the trouble of burning down some first-generation 5G networking infrastructure, 5G is dangerous. 5G will kill us all, sterilise our reproductive systems, and it's also all about global-scale digital mind control.

Let's bypass some of the subtleties and not-so-subtleties here and boil all of these questions about the safety of 5G down to one simple one: can 5G kill you? It turns out that to answer that, you have set aside your Twitter feed and do some real scientific research.

An important thing to remember when considering the risks of 5G, both real and imagined, is that we're all constantly taking a bath in an invisible mixture of radio waves. Yes, you, yourself, reading this. Just take a moment and think about your phone, your computer, your Wi-Fi, your Bluetooth headphones, your TV, your smart speaker, and remember that they are all, already, radiating. All around you, all the time. Remember that your phone can pick up a signal inside your home, so the signals are coming through your walls. Remember that your neighbour probably has Wi-Fi and lots of devices too. Then there are terrestrial TV signals, traditional music radio signals, police and emergency services walkie-talkies, satellite TV and communications signals radiating down from space, including GPS. Oh, and I've not mentioned radar signals from aircraft, radar signals from weather scanning equipment and probably a ton of other radio signals. Remember I said the radio spectrum was so full of people using it that we had to clear out some space for 5G? Yup. And before I move on, let me just point out that there are also plenty of natural

You can't hide from 5G...

sources of radio frequency radiation, including exotic ones that come from sources like the echo of the Big Bang . . . which just happens to be in the microwave band.

Why am I saying this? It's to point out that whether you know it or not, you are being exposed to all sorts of natural and artificial radio signals all the time. It's too late to pop a tin-foil hat on your head.

Really dangerous waves

The next important thing to know is that there really are some forms of electromagnetic radiation that are dangerous. These are radiations that are 'ionising', which means that when a wave hits an atom of something, whether that atom is in a rock or in a cell in your brain, it can knock off an electron from the atom and turn it into an ion. It's a very natural process, and when you make it happen artificially, it can be very useful. But you really don't want it to happen in certain places: ionisation in a rock isn't something you probably need to be concerned about, but lots of ionising waves going through your brain or hitting your skin, for example, would damage the cells.

So what are these ionising electromagnetic radiations? Gamma rays are a classic example, and one that the Incredible Hulk is deeply familiar with. They also *sound* dangerous, and are the sort of thing that nuclear weapons shoot out, aren't they? (Yes!) X-rays are another example. X-rays are incredibly useful when you've slipped when rollerblading and broken your elbow, but X-ray technicians wear protective lead aprons for a reason. Lastly, certain UV rays are dangerously ionising. You probably know that some UV light is something to be wary of, thanks to skin damage and the heightened risk of skin cancers.

But apart from these particular electromagnetic radiations, nothing else is classified as 'ionising'. Not visible light, obviously. Not infrared. Not microwaves. Not TV signals. Not 3G cell phone radio signals. Not 4G radio signals. And not 5G radio signals. Those new 5G millimetre wave signals? Nope. Not ionising.

Enthusiastic sceptics will point out that some of these non-ionising radio signals actually are dangerous in their own way. For example, your microwave oven has all sorts of interlocks and filters to prevent the microwave radio signals from getting out and cooking you instead of merely warming up a bowl of soup. A really bright light can damage your vision permanently. Meanwhile, some lasers or bright infrared sources can actually burn your skin. And yes, all of this is true. But in all these examples, the electromagnetic signals causing the trouble are delivered in really high-power ways, and often the problem comes when you direct the waves very precisely on to people. The doses and power levels that cause damage under these circumstances are very well known; to put it in context, they are often thousands if not millions or billions of times more powerful than the unfocused radio signals put out by your phone or Wi-Fi router, or showering down on you from GPS satellites and so on.

Scientists and medical experts from all sorts of fields have spent a long time looking at the risks and dangers of all sorts of radio frequency radiations, and their general opinion is that there is no proof of any sort of risk associated with using radio devices like smartphones, or even merely existing in their radio field. The World Health Organization said way back in 2014 that there was no evidence of risk from mobile phone use, and for the most part 5G will use similar radio signals to the phone frequencies studied. The UK government says that even if you stand near 5G towers, the radiation levels are far below accepted guidelines. Other governments around the world have come to the same conclusion, even for millimetre wave radio signals.

Some will scoff at this point, and say, 'You can't trust a scientist!' I can't argue with you there, but I will point out that scientists and engineers also designed your phone, your TV, your dishwasher and your car's safety system, and they were involved in designing, testing and improving nearly every bit of medical care you've ever experienced.

Are scientists and doctors ever wrong? Sure. All the time. It's actually a part of the scientific learning process, and don't let a social media bully or a misinformed politician tell you otherwise! But over time, the level of expertise among scientists grows, and the understanding of the entire scientific community improves – and it's the whole community that matters, not one or two outspoken scientists who buck the trend. Because just as with any community of supposedly similar people, there are always firebrands. It just sounds extra convincing if you're a firebrand who can speak maths!

So, the important point is this: on the whole, thanks to decades of research and experiment, the scientific conclusions point to no proven risk to health from 5G radio. Here are the World Health Organization's words on the matter, as of late 2020:

> To date, and after much research performed, no adverse health effect has been causally linked with exposure to wireless technologies. Health-related conclusions are drawn from studies performed across the entire radio spectrum but, so far, only a few studies have been carried out at the frequencies to be used by 5G.

Tissue heating is the main mechanism of interaction between radiofrequency fields and the human body. Radiofrequency exposure levels from current technologies result in negligible temperature rise in the human body.

As the frequency increases, there is less penetration into the body tissues and absorption of the energy becomes more confined to the surface of the body (skin and eye). Provided that the overall exposure remains below international guidelines, no consequences for public health are anticipated.

The coronavirus conundrum

This conspiracy theory is, to a scientific mind, a proper side-splitter. The theory alleges that there is no such illness as COVID-19, and that actually it's all a cover-up. What it's actually a cover-up *for* is not clearly described, but the allegations cover everything from mind control to 4G and 5G causing cancer. Sometimes the suggestion is that 5G is itself spreading the novel coronavirus strain. Oh my.

This last bit is easy to set straight. A biological virus is a microscopic bundle of atoms that can invade living cells in plants or animals, where it then settles down and gets on with the business of making copies of itself. The cell can be damaged by this, but the virus doesn't care. As far as the virus is concerned, its only mission is to replicate. The common cold is a virus. HIV is a virus. Herpes is a virus. COVID-19 is a virus.

Remember my physics chapter? Radio signals are waves of electric and magnetic energy that shoot through space at the speed of light. They are generated naturally and by human-made electronics. They are *not* a virus. Millimetre wave signals, which are radio waves, are not a virus. Neither are other 5G, 4G and 3G signals, Bluetooth and so on. Radio signals are not viruses in exactly the same way that a red London bus is not a virus.

A really persistent conspiracy theorist may chime in at this point and say, 'Ah, but actually coronavirus is a cover-up operation for the damage that 5G is already doing!' And to them I would say, 'Ah, but what if that's what "they" want you to think? What if this cover-up is itself is a cover-up? The truth is that really it's all about the aliens . . .'

*

In conclusion, can 5G kill you? If a 5G delivery drone malfunctioned and lost power and dropped out of the sky on to your head, then yes. Maybe. Other than this, no, probably not.

And here's the inevitable caveat. I am a scientist, but am I a medical expert? No. It's important to point this out, and it's really not a writer's responsibility to make decisions about your health. But research by actual *medical experts into the health risks of radio signals of all sorts, including 5G wavelengths and millimetre waves, is ongoing around the world. It's happening right now, and so far that research is giving 5G a thumbs-up.*

Chapter 7

You Know What's Coming . . .
5-and-a-Bit G!

5G is only getting started right now, with different phone networks rolling out 5G of different types and radio frequencies at different times in different places. It's really not a coordinated effort, and the same thing happened when 4G arrived. 4G evolved over time, getting faster and more reliable and so on as enhanced technologies were turned on and people like you and me bought new phones to take advantage of the newer systems. From all this history, we can expect that 5G itself will change over the next decade.

5G for early adopters

Some of the first 5G to arrive will overlap in a number of important ways with 4G. Technically, some of the 5G bits your phone connects to will, behind the scenes, be plugged into some of the same radio and computer boxes that 4G LTE systems plug into – the infrastructure that makes the network work. This is a good idea for the networks because it costs them less, and it gets the whole 5G ball rolling with us, the general public. This kind of 5G is called 'non stand-alone', and while technically it *can* bring some of the advantages of 5G, like speed and lower latency, it can't guarantee them. Which means, yup, you've guessed it: it may not be much better than 4G.

Proof positive of the fact that 5G is arriving at different times and different speeds in different places was delivered by Apple in 2020, when it only released the 5G-capable iPhone 12 with mmW tech for US-based 5G networks – which was only switched on in a limited number of cities at the time! People in other countries didn't get to enjoy super-speedy iPhone 12s, even though mmW services were going live.

But as 5G networks expand across the world, more radio spectrum space is freed up for them and new technologies like mmW go live in more places, 5G *will* begin to deliver on its promise. This phase, where the 5G network relies on its own infrastructure, is the 'stand-alone' phase. It's the exciting bit, and it may take a few years to get there, but by the middle of the 2020s, it will be in full swing.

5G gets into the swing of it

In the late 2020s, 5G is expected to evolve even more. This is when the massive Internet of Things boost is expected. 5G network speeds will probably be turned up to volume 11 and all sorts of new 5G network uses like vehicle-to-vehicle communications will happen.

Marketing and the bloodthirsty hunt for profit isn't going to go away, of course, and I can imagine that before the decade is out, there will be networks advertising '5G+' networks and phones and plenty of other gizmos. They'll say they are 'faster than 5G' or some such. And they probably *will* be faster than the 5G systems that are going online as this book is being written.

*

The thing is, no one can exactly *say how 5G will evolve: these decisions are still being made, and governments aren't even done with clearing out radio spectrum for the 5G available right now, let alone the 5G of 2029.*

This book is all about 5G, which is even now rolling out across the world as a revolutionary new technology, but did you know that the early ideas for 5G began to be dreamed up around 2008? Yup. It took ten years for those ideas to become reality! For about ten years before 5G, 4G was the big news, of course. And for about ten years before 4G, 3G was all the rage. Which means that right now, *in boardrooms and committee meetings and in oh so many tedious memos and financial spreadsheets (and much more interestingly in scientists' and engineers' minds and on their computers), some even newer ideas are being crafted for the yet more revolutionary and blood-tinglingly amazing technology that will arrive in about ten years' time:*

6G!

6G, the fantasy dream

One of the first fledgling commercial 5G networks was launched in 2019 by NTT DOCOMO, the Japanese cell phone giant (although other networks in other countries launched and test-launched 5G earlier). In 1979, NTT also launched the first commercial cellular network, and they had one of the first 3G networks, so maybe when these people talk about the next G, we should listen. They've already released a white paper that roughly describes what 6G should be like. It is pretty amazing, and it all spins around one word: extreme.

NTT says that where 5G has good connection reliability, 6G should have *extreme* reliability. While 5G has high-speed data rates, 6G should have *extreme* data rates. Where 5G has ultra-low latency, in 6G we should expect *extreme* low latency. Basically, NTT says 6G should do everything that 5G does but better and faster, including having minimum data speeds of 100 gigabits per second. That is, and there's no other way to say it, ludicrous speed.*

What will this amazing tech allow? NTT's experts hope it could help solve social problems, particularly erasing the barriers between urban and rural areas. It could enable more communication between humans and things, revolutionising entertainment and enterprise services. It should allow communication everywhere, even in space and in the middle of the ocean. Somewhat amazingly, NTT even suggests that we'll see 6G 'cyber-physical fusion', which sounds bonkers, but it means that micro-scale

---

*At 100 gigabit speeds, a 4.5 megabyte PDF file containing Shakespeare's entire works will download in 0.00036 seconds. If you blinked while reading those words, a 6G phone could complete around 1,000 complete Shakespeare downloads *while your eyes were closed.* Put it another way: human DNA takes about 1.5 GB to write down, so at 6G speeds you could download the entire code that makes you *you* in around a tenth of a second.

wearable 6G tech could be possible, and that part of our lives will be lived in a type of cyberspace.

The clever boffins at Samsung also have their own 6G white paper. In many ways, it lines up with NTT's vision, but it also suggests some truly wild stuff, talking of mobile holograms and extreme XR that can create 'digital replicas' of things. It's fantasy, of course, but highly informed, highbrow fantasy, plus it comes from the minds of people who will actually be building this stuff, so it's worth taking note.

From these two sources alone, plus extrapolating the trends we've already seen for 4G and 5G, it seems that 6G will bring extraordinary data speeds and mind-boggling levels of connectivity, leading to all sorts of crazy innovations off the back of it: high-precision *indoor* GPS navigation, for one pretty fascinating example.

6G, the cold, hard reality

The 6G vision is beginning to take shape, but the question is: how will 6G happen in reality? What technology will actually make it work? The cool bit about the answer to this is, er . . . we don't know, exactly. But we do have some ideas.

Just as 5G saw an expansion of the bandwidth used for mobile broadband, we can expect 6G to use more radio spectrum and find clever new ways to maximise the radio bandwidth it's using. Some of the technologies used for 6G may even merge with next-generation Wi-Fi, not least because it's simpler and cheaper to build devices that use just one set of chips. Your 4G phone, for example, has radio antennas optimised for 4G, 3G, Wi-Fi, Bluetooth and so on. Your 5G phone will have these, and will also need mmW antennas. Your 6G phone? Well . . . Will it need so many? Or more? Terahertz radio frequencies might be used for 6G, which means radio waves quite close to infrared light, and these frequencies could mean ultra-precise beamforming (remember, this is the way that signals are pointed carefully in certain directions). The wavelength of terahertz signals could even involve us using weird 'metamaterial' surfaces on walls and so on to bounce network signals where they need to go – so the decor of your room may help your network connectivity.

6G may even blur what we think of as 'being connected'. Mobile broadband, Wi-Fi and maybe even space-based Internet, even now being launched by companies like SpaceX, could all fuse into one experience. You'd basically turn your device on and it would just be 'on', connected, anywhere and everywhere. What this will do for the future of Internet Service Providers and mobile phone networks is unpredictable (though you can bet your bottom dollar that these companies will still love to take that bottom dollar from you, along with many more, in exchange for the privilege of letting you connect to their 6G networks).

6G may even include some open source principles, with less company-owned, trademarked and patent-protected hardware and software inside the cell phone radio masts and server centres, and maybe even in the chips on 6G phones, and instead more 'open' radio networks, making it easier to innovate wholly new ways to use the tech.

If 5G is being pushed as the best thing since sliced bread, 6G will just have to be 'better than sliced bread'.

*

Whatever ends up happening will end up happening. It's going to be fascinating to see what impact

6G will have on the world, because while 5G is busy bringing a sort of 'home broadband' experience to mobile devices (and who knows exactly what amazing changes that will mean in the end), 6G's speeds and extreme connectivity will surpass 5G by light years.

Epilogue

And, lo, at the closing of the great launch day for 5G, the stage technicians were toiling hard amid the wispy remains of the special-effects smoke. With great brooms they swept up the mess, including snack wrappers that the bloody musicians always cast upon the floor. Then with careful hands they did coil cables and pack away the speakers and the lasers in heavy cases with much squeaking of strong polystyrene. The pyro team did clean and oil the cannons before rolling them into giant storage boxes with great rumbling sounds.

Finally the last crate was heaved into the storage room, and the gates of security were closed with a clang. Quietly the technicians left, all but one, murmuring among themselves about beer's blessed rejuvenating powers.

The chief technician looked out across the digital wilderness, then speedily scrolled through some info shown upon a shiny new 5G phone. This technician did quickly write upon a label, and slapped it across the lock of the great chamber. It read 'Ready for 6G'. Then, with a 'click!' that echoed mightily, they turned off the light . . .